A READY FOR WORK
RECORD BOOK

MATHS FOR LIFE

G. LEWIS

COLLINS
GLASGOW & LONDON

Contents

Whole Numbers—Addition

1 $\begin{array}{r} 3\,5 \\ +\ 3 \\ \hline \end{array}$ **2** $\begin{array}{r} 2\,3 \\ +4\,4 \\ \hline \end{array}$ **3** $\begin{array}{r} 4\,6 \\ +\ 5 \\ \hline \end{array}$ **4** $\begin{array}{r} 3\,7 \\ +1\,4 \\ \hline \end{array}$

5 $8\,2 + 7$ **6** $\begin{array}{r} 5\,6 \\ +\ 8 \\ \hline \end{array}$ **7** $\begin{array}{r} 6\,5 \\ +3\,6 \\ \hline \end{array}$

8 $\begin{array}{r} 2\,3\,1 \\ +\ 4 \\ \hline \end{array}$ **9** $\begin{array}{r} 4\,2\,6 \\ +\ 9 \\ \hline \end{array}$ **10** $374 + 18$

11 $\begin{array}{r} 5\,5\,3 \\ +\ 6\,8 \\ \hline \end{array}$ **12** $\begin{array}{r} 1\,5\,2 \\ +1\,2\,7 \\ \hline \end{array}$ **13** $\begin{array}{r} 5\,8\,2 \\ +3\,2\,9 \\ \hline \end{array}$

14 $\begin{array}{r} 7\,6\,3 \\ 2\,2 \\ +\ 9 \\ \hline \end{array}$ **15** $678 + 13 + 22$

16 $\begin{array}{r} 1\,6\,4 \\ +4\,5\,6 \\ \hline \end{array}$ **17** $\begin{array}{r} 2\,5\,6 \\ 2\,1\,4 \\ +4\,2\,3 \\ \hline \end{array}$ **18** $\begin{array}{r} 3\,1\,5 \\ 2\,4\,8 \\ +9\,7 \\ \hline \end{array}$

19 $\begin{array}{r} 4\,7\,8 \\ 2\,2\,3 \\ +3\,1\,9 \\ \hline \end{array}$ **20** $219 + 36 + 23 + 8$

21 $\begin{array}{r} 1\,0\,4\,9 \\ 1\,6 \\ +1\,8\,4 \\ \hline \end{array}$ **22** $\begin{array}{r} 2\,6\,7\,6 \\ 1\,0\,8 \\ +7\,2 \\ \hline \end{array}$ **23** $\begin{array}{r} 4\,5\,6\,9 \\ 2\,0\,0\,3 \\ +5\,2\,6\,9 \\ \hline \end{array}$

24 $1069 + 17 + 346 + 2$

25 In an election, the Labour candidate received 25 405 votes, the Conservative 13 244 and the Liberal 8206. How many votes were cast altogether?

26 A trolley has an unladen weight of 30 kg. It is loaded with three cases weighing 10 kg, 5 kg and 7 kg. The axles can safely support a maximum load of 50 kg. Is the trolley safe? _____

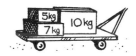

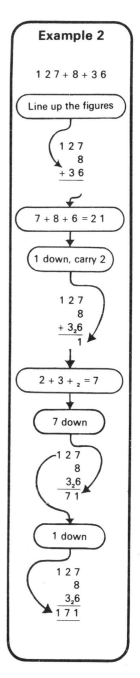

Whole Numbers—Subtraction

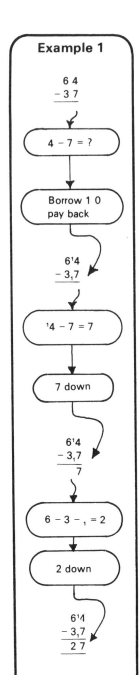

Example 1

```
  6 4
− 3 7
```

4 − 7 = ?

Borrow 1 0
pay back

```
  6¹4
− 3,7
```

¹4 − 7 = 7

7 down

```
  6¹4
− 3,7
    7
```

6 − 3 − , = 2

2 down

```
  6¹4
− 3,7
   2 7
```

1
```
 6 7
− 5
```

2
```
 9 5
−1 4
```

3
```
 6 2
−1 3
```

4 8 4 − 2 7 _____

5
```
 5 3
− 7
```

6
```
  7 3
−6 5
```

7
```
4 3 6
  − 3
```

8
```
3 5 2
  − 3
```

9 5 2 4 − 6 _____

10
```
1 3 5
 −2 4
```

11
```
6 2 8
 −3 7
```

12 2 4 6 − 2 8 _____

13
```
7 2 1
 −3 6
```

14
```
5 3 2
 −3 4
```

15
```
7 4 5
 −6 6
```

16
```
3 7 6
 −8 7
```

17
```
 7 6 4
−3 5 2
```

18
```
 4 5 4
−1 2 5
```

19 9 1 1 − 6 2 2 _____

20
```
 6 3 2
−3 5 8
```

21
```
2 0 6 7
  −3 3
```

22
```
1 6 8 5
 −3 2 4
```

23
```
4 5 2 1
 −1 1 9
```

24
```
 2 6 7 9
−1 2 4 6
```

25
```
 8 9 7 8
−3 6 9 4
```

26
```
 3 0 6 5
−2 9 9 9
```

27
8 7 − 2 3
2 3 to 3 0 is _____
3 0 to 8 0 is _____
8 0 to 8 7 is _____
Total _____

28
1 5 0 − 4 9
4 9 to 5 0 is _____
5 0 to 1 0 0 is _____
1 0 0 to 1 5 0 is _____
Total _____

Example 2

```
  4 0 6 5
− 2 3 8 7
```

```
  4 0 6¹5
− 2 3 8,7
        8
```

```
  4 0¹6 5
− 2 3,8,7
       7 8
```

```
  4¹0 6 5
− 2,3,8 7
     6 7 8
```

```
  4 0 6 5
− 2,3 8 7
  1 6 7 8
```

Example 3

```
 3 6 2
 − 7 6
```

Shopkeeper method
```
 7 6 to   8 0 is     4
 8 0 to 1 0 0 is   2 0
1 0 0 to 3 0 0 is 2 0 0
3 0 0 to 3 6 2 is  6 2
           Total 2 8 6
```

---

29 A weighing machine gives a reading of 4 kg when it is unloaded. A container, weighing 1 kg, is put onto the machine and filled with sand. The machine then reads 23 kg. What is the weight of the sand? _____

Whole Numbers—Multiplication

Example 1

13×7

$$
\begin{array}{r}
1\,3 \\
\times 7 \\
\hline
\end{array}
$$

Multiply from the right

$7 \times 3 = 21$

1 down, carry 2

$$
\begin{array}{r}
1\,3 \\
\times 7 \\
\hline
{}_2 1
\end{array}
$$

$7 \times 1 + {}_2 = 7 + {}_2$

9 down

$$
\begin{array}{r}
1\,3 \\
\times 7 \\
\hline
9\,{}_2 1
\end{array}
$$

Example 3

$$
\begin{array}{r}
3\,4\,7 \\
\times 2\,8 \\
\hline
\end{array}
$$

Multiply by units first

$$
\begin{array}{r}
3\,4\,7 \\
\times 2\,8 \\
\hline
2\,7\,{}_3 7\,{}_5 6
\end{array}
$$

Nought down, multiply by 2

$$
\begin{array}{r}
3\,4\,7 \\
\times 2\,8 \\
\hline
2\,7\,7\,6 \\
6\,9\,4\,0
\end{array}
$$

Then add

$$
\begin{array}{r}
2\,7\,7\,6 \\
+\,6\,9\,4\,0 \\
\hline
9\,7\,1\,6
\end{array}
$$

1 $\begin{array}{r} 1\,2 \\ \times 4 \\ \hline \end{array}$ **2** $\begin{array}{r} 3\,2 \\ \times 3 \\ \hline \end{array}$ **3** $\begin{array}{r} 1\,6 \\ \times 2 \\ \hline \end{array}$

4 $\begin{array}{r} 1\,8 \\ \times 5 \\ \hline \end{array}$ **5** $\begin{array}{r} 2\,3 \\ \times 4 \\ \hline \end{array}$ **6** $\begin{array}{r} 2\,7 \\ \times 5 \\ \hline \end{array}$

7 $\begin{array}{r} 3\,5 \\ \times 6 \\ \hline \end{array}$ **8** $\begin{array}{r} 2\,4 \\ \times 9 \\ \hline \end{array}$ **9** $\begin{array}{r} 1\,5\,8 \\ \times 7 \\ \hline \end{array}$

10 $\begin{array}{r} 1\,3\,2 \\ \times 2 \\ \hline \end{array}$ **11** $\begin{array}{r} 2\,1\,6 \\ \times 3 \\ \hline \end{array}$ **12** $\begin{array}{r} 1\,2\,4 \\ \times 4 \\ \hline \end{array}$

13 $\begin{array}{r} 1\,4\,8 \\ \times 8 \\ \hline \end{array}$ **14** $\begin{array}{r} 3\,5\,9 \\ \times 2\,5 \\ \hline \end{array}$ **15** $\begin{array}{r} 4\,0\,6 \\ \times 1\,7 \\ \hline \end{array}$

16 $\begin{array}{r} 2\,4 \\ \times 1\,2 \\ \hline \end{array}$ **17** $\begin{array}{r} 3\,8 \\ \times 2\,6 \\ \hline \end{array}$ **18** $\begin{array}{r} 1\,2\,7 \\ \times 1\,3 \\ \hline \end{array}$

19 $\begin{array}{r} 3\,0\,5\,2 \\ \times 1\,4 \\ \hline \end{array}$ **20** $\begin{array}{r} 4\,1\,5\,8 \\ \times 2\,5 \\ \hline \end{array}$ **21** $\begin{array}{r} 6\,3\,2\,7 \\ \times 6\,8 \\ \hline \end{array}$

Example 2

105×5

$$
\begin{array}{r}
1\,0\,5 \\
\times 5 \\
\hline
\end{array}
$$

$5 \times 5 = 25$

5 down, carry 2

$$
\begin{array}{r}
1\,0\,5 \\
\times 5 \\
\hline
{}_2 5
\end{array}
$$

$5 \times 0 + {}_2 = 0 + 2$

2 down

$$
\begin{array}{r}
1\,0\,5 \\
\times 5 \\
\hline
2\,{}_2 5
\end{array}
$$

$5 \times 1 = 5$

5 down

$$
\begin{array}{r}
1\,0\,5 \\
\times 5 \\
\hline
5\,2\,{}_2 5
\end{array}
$$

22 A factory uses 15 tonnes of sand per day. How much does it use in a 20 working-day month?

23 If each pupil in a school uses 8 notebooks a year, how many will 843 pupils use?

Whole Numbers—Division

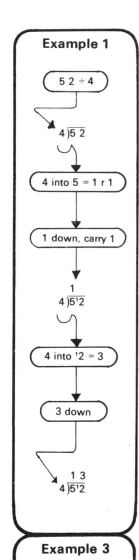

Example 1

$$5\,2 \div 4$$

$$4\,\overline{)5\,2}$$

4 into 5 = 1 r 1

1 down, carry 1

$$\overset{1}{\underset{4\,\overline{)5^1 2}}{}}$$

4 into ¹2 = 3

3 down

$$\overset{1\ 3}{\underset{4\,\overline{)5^1 2}}{}}$$

1 $2\,\overline{)4\,2}$ **2** $3\,\overline{)3\,9}$ **3** $5\,\overline{)6\,5}$ **4** $9\,8 \div 7$

5 $6\,\overline{)8\,4}$ **6** $4\,\overline{)8\,0\,8}$ **7** $3\,\overline{)1\,2\,9}$

8 $6\,\overline{)1\,8\,6}$ **9** $6\,5\,5 \div 5$ **10** $8\,\overline{)9\,2\,8}$

11 $7\,\overline{)4\,1\,3}$ **12** $9\,\overline{)7\,0\,2}$ **13** $1\,2\,\overline{)2\,5\,2}$

14 $1\,6\,\overline{)4\,9\,6}$ **15** $4\,2\,9 \div 1\,3$ **16** $1\,9\,\overline{)4\,3\,7}$

17 $2\,4\,\overline{)8\,1\,6}$ **18** $5\,9\,\overline{)7\,0\,8}$ **19** $3\,5\,\overline{)4\,5\,5}$

20 $2\,8\,\overline{)8\,1\,2}$ **21** $3\,6\,\overline{)6\,1\,2}$ **22** $2\,0\,\overline{)8\,4\,0}$

Example 3

$$\begin{array}{r} 2\,4 \\ 2\,3\,\overline{)5\,5\,2} \\ -4\,6 \\ \hline 9\,2 \\ -9\,2 \\ \hline 0\,0 \end{array}$$

23 $1\,5\,\overline{)1\,2\,9\,0}$ **24** $1\,3\,\overline{)1\,2\,4\,8}$ **25** $1\,6\,\overline{)1\,5\,3\,6}$

Example 2

$$1\,6\,2 \div 6$$

$$6\,\overline{)1\,6\,2}$$

6 into 1 ?

Keep on

6 into 1 6 = 2 r 4

2 down, carry 4

$$\overset{2}{\underset{6\,\overline{)1\ 6^4 2}}{}}$$

6 into ⁴2 = 7

7 down

$$\overset{2\ 7}{\underset{6\,\overline{)1\ 6^4 2}}{}}$$

26 A shop assistant goes to the bank to withdraw some cash for the till. She asks for £65 to be paid in five pound notes. How many does she get? _____

27 A family hire a television and pay £156 a year. How much is this per week? _____

Whole Numbers—Averages

The **average** or **mean** of a set of numbers is the total divided by how many numbers there are.

Example 1

Find the average of:

3, 2, 5, 6 and 4

↓

(Find the total)

3 — ¹
2 — ²
5 — ³
6 — ⁴
4 — ⁵

→ 2 0

(Divide by number)

(2 0 ÷ 5 = 4)

Average = 4

Example 2

Find the average of:

5 6, 6 2, 8 6 and 5 2

↓

(Find the total)

↓

(Divide by number)

$$\begin{array}{r} 6\ 4 \\ 4\overline{)2\ 5\ 6} \end{array}$$

Average = 6 4

Example 3

I travel 250 miles in 5 hours.
My average speed is
2 5 0 ÷ 5 = 5 0 mph

---

1 Find the average of:

   a 4, 3, 5, 4, 7, 7

   Average is _____

   b 9, 8, 11, 10, 7

   Average is _____

   c 21, 22, 24, 22, 27, 22

   Average is _____

   d 17, 15, 11, 19, 24, 29, 18

   Average is _____

   e 112, 153, 167

   Average is _____

2 Ten screwdrivers were measured (to the nearest mm).
They were: 115, 114, 116, 117, 116, 115, 117, 115, 117 and 118.
What is the average length?

_____

3 Find the average speed if I travel:

   a 240 km in 6 hours _____ km/h
   b 70 miles in 2 hours _____ mph
   c 480 km in 6 hours _____ km/h
   d 168 km in 3 hours 30 minutes _____ km/h

Fractions—Equivalent Fractions

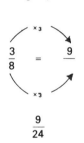
1 Complete these different ways of writing the same fraction, by putting the correct numbers in the spaces.

$$\frac{1}{2} \quad = \quad \frac{2}{4} \quad = \quad \frac{3}{\rule{1em}{0.4pt}} \quad = \quad \frac{\rule{1em}{0.4pt}}{16} \quad = \quad \frac{4}{8}$$

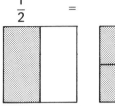 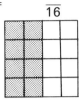 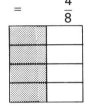

2 What fraction of each shape is shaded?

 _____   _____ 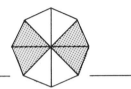 _____

3 Find then shade in these equivalent fractions.

a $\dfrac{1}{3} = \dfrac{\rule{1em}{0.4pt}}{6}$ **b** $\dfrac{3}{8} = \dfrac{\rule{1em}{0.4pt}}{16}$

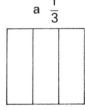 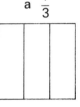 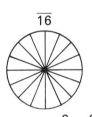

4 $\dfrac{1}{4} = \dfrac{\rule{1em}{0.4pt}}{8}$

5 $\dfrac{2}{5} = \dfrac{4}{\rule{1em}{0.4pt}}$

6 $\dfrac{3}{10} = \dfrac{\rule{1em}{0.4pt}}{100}$

7 $\dfrac{9}{10} = \dfrac{90}{\rule{1em}{0.4pt}}$

8 $\dfrac{1}{2} = \dfrac{\rule{1em}{0.4pt}}{100}$

9 $\dfrac{1}{10} = \dfrac{2}{\rule{1em}{0.4pt}}$

10 $\dfrac{4}{5} = \dfrac{\rule{1em}{0.4pt}}{20}$

11 $\dfrac{3}{4} = \dfrac{9}{\rule{1em}{0.4pt}}$

12 $\dfrac{1}{3} = \dfrac{\rule{1em}{0.4pt}}{12}$

13 $\dfrac{3}{8} = \dfrac{9}{\rule{1em}{0.4pt}}$

14 $\dfrac{5}{8} = \dfrac{\rule{1em}{0.4pt}}{16}$

15 $\dfrac{7}{10} = \dfrac{35}{\rule{1em}{0.4pt}}$

16 $\dfrac{2}{3} = \dfrac{\rule{1em}{0.4pt}}{24}$

17 $\dfrac{7}{16} = \dfrac{28}{\rule{1em}{0.4pt}}$

18 $\dfrac{9}{10} = \dfrac{\rule{1em}{0.4pt}}{120}$

19 $\dfrac{6}{14} = \dfrac{18}{\rule{1em}{0.4pt}}$

20 In a $\dfrac{1}{96}$th model of an aircraft, a part is 5 cm long. How long is this part in the real aircraft? $\dfrac{1}{96} = \dfrac{5}{?}$ _____

21 In a survey of 180 people, two-thirds preferred DAZZLE toothpaste. How many people is this? $\dfrac{2}{3} = \dfrac{?}{180}$ _____

Fractions—Improper Fractions

Improper fractions are fractions that are 'top heavy'. This means that the number on the top, the NUMERATOR, is bigger than the number on the bottom, the DENOMINATOR. They can be converted to give a whole number and fraction.

Example 1

Three halves

$$\frac{3}{2} = \frac{2+1}{2}$$

$$= \frac{2}{2} + \frac{1}{2} = 1 + \frac{1}{2}$$

$$= 1\frac{1}{2}$$

Example 2

$$\frac{27}{8}$$

$$= \frac{8+8+8+3}{8}$$

$$= 3\frac{3}{8}$$

Example 3

We can also change them back again . . .

$$2\frac{2}{5}$$

$$= \frac{5+5+2}{5}$$

$$= \frac{12}{5}$$

1 Convert these improper fractions to mixed numbers:

a $\frac{5}{2}$ _____   **b** $\frac{7}{4}$ _____   **c** $\frac{11}{4}$ _____

d $\frac{6}{5}$ _____   **e** $\frac{22}{5}$ _____   **f** $\frac{13}{8}$ _____

g $\frac{5}{3}$ _____   **h** $\frac{15}{3}$ _____   **i** $\frac{27}{10}$ _____

j $\frac{50}{12}$ _____   **k** $\frac{314}{100}$ _____

2 Convert these mixed numbers to improper fractions:

a $2\frac{1}{2}$ _____   **b** $1\frac{1}{4}$ _____   **c** $3\frac{3}{4}$ _____

d $2\frac{1}{3}$ _____   **e** $3\frac{2}{3}$ _____   **f** $3\frac{2}{5}$ _____

g $4\frac{5}{8}$ _____   **h** $7\frac{3}{8}$ _____   **i** $9\frac{3}{10}$ _____

3 What is the numerator in two-thirds? _____

4 What is the denominator in five-eighths? _____

5 The counting numbers are 1, 2, 3, . . . What denominator do they have? _____

6 Ages are often written as mixed numbers. A girl who is 13 years and 5 months old would write her age as $13\frac{5}{12}$.

Complete this table.

Years and Months		Months	Mixed number
3	7	43	$3\frac{7}{12}$
8	2		
			$14\frac{5}{12}$
		204	

9

Fractions—Cancelling

Some fractions can be simplified by dividing the numerator and the denominator by the same number. This operation is called **cancelling**.

Example 1

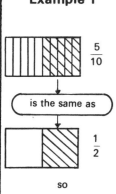

$\frac{5}{10}$

is the same as

$\frac{1}{2}$

so

Divide top and bottom by same number

$\frac{5}{10} = \frac{1}{?}$

÷ by 5

$\frac{5}{10} = \frac{1}{2}$

÷ by 5

1
   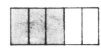

$\frac{6}{10}$ $=$ $—$

2 $\frac{2}{10} \begin{smallmatrix} \div 2 \\ \\ \div 2 \end{smallmatrix} = \frac{1}{2}$

3 $\frac{4}{8} \begin{smallmatrix} \div 4 \\ \\ \div 4 \end{smallmatrix} = \frac{1}{ }$

4 $\frac{10}{16} \begin{smallmatrix} \div 2 \\ \\ \div 2 \end{smallmatrix} = \frac{ }{8}$

5 $\frac{8}{32} = \frac{ }{8}$

6 $\frac{24}{36} = \frac{ }{12}$

7 $\frac{12}{32} =$

8 $\frac{15}{60} =$

9 $\frac{3}{15} =$

10 $\frac{14}{21} =$

11 $\frac{15}{25} =$

12 $\frac{100}{250} =$

13 $\frac{125}{200} =$

14 $\frac{210}{500} =$

15 $\frac{126}{200} =$

16 $\frac{12}{9} =$

17 $\frac{18}{6} =$

18 $\frac{36}{8} =$

19 $\frac{60}{24} =$

20 $\frac{144}{96} =$

21 In a school, 600 pupils are boys and 650 are girls. What fraction of the pupils are boys?

_____

22 In a car factory 60 workers produce 1500 cars a week. How many on average does each worker produce per week? _____

Example 2

$\frac{12}{36}$

↓

Divide by 2

↓

$\frac{12}{36} = \frac{6}{18}$

↓

Keep dividing as far as you can

$\frac{6}{18}$

↓

Divide by 3

↓

$\frac{6}{18} = \frac{2}{6}$

↓

Divide by 2

↓

$\frac{2}{6} = \frac{1}{3}$

Example 3

$\frac{220}{80}$

↓

Divide by 10

$\frac{220}{80} = \frac{22}{8}$

↓

Divide by 2

$\frac{22}{8} = \frac{11}{4}$

$\frac{11}{4} = \frac{4+4+3}{4}$

$= 2\frac{3}{4}$

Fractions—Addition and Subtraction

To add or subtract fractions, find equivalent fractions with the same denominator.

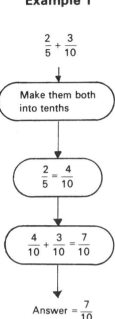

Example 1

$$\frac{2}{5} + \frac{3}{10}$$

Make them both into tenths

$$\frac{2}{5} = \frac{4}{10}$$

$$\frac{4}{10} + \frac{3}{10} = \frac{7}{10}$$

Answer $= \frac{7}{10}$

Example 2

$$\frac{1}{3} + \frac{3}{4}$$

Make them into twelfths

$$\frac{1}{3} = \frac{4}{12}$$

$$\frac{3}{4} = \frac{9}{12}$$

$$\frac{4}{12} + \frac{9}{12} = \frac{13}{12}$$

$$\frac{13}{12} = \frac{12 + 1}{12}$$

$$= 1\frac{1}{12}$$

Example 3

$$2\frac{5}{8} - 1\frac{1}{3}$$

Make them into improper fractions

$$2\frac{5}{8} = \frac{21}{8}$$

$$1\frac{1}{3} = \frac{4}{3}$$

Make them into equivalent fractions

$$\frac{21}{8} = \frac{63}{24}$$

$$\frac{4}{3} = \frac{32}{24}$$

$$\frac{63}{24} - \frac{32}{24} = \frac{31}{24}$$

$$\frac{31}{24} = 1\frac{7}{24}$$

1 $\frac{1}{4} + \frac{1}{4}$

2 $\frac{3}{4} - \frac{1}{4}$

3 $\frac{1}{5} + \frac{7}{10}$

4 $\frac{7}{8} - \frac{3}{4}$

5 $\frac{1}{2} + \frac{3}{8}$

6 $\frac{11}{12} - \frac{3}{4}$

7 $\frac{1}{2} + \frac{1}{3}$

8 $\frac{2}{3} - \frac{1}{2}$

9 $\frac{1}{3} + \frac{1}{4}$

10 $\frac{1}{2} - \frac{1}{6}$

11 $\frac{2}{5} + \frac{1}{2}$

12 $\frac{5}{6} - \frac{1}{4}$

13 $\frac{1}{4} + \frac{3}{5}$

14 $\frac{3}{8} - \frac{1}{3}$

15 $\frac{5}{12} + \frac{1}{3}$

16 $\frac{4}{7} - \frac{2}{5}$

17 $\frac{3}{8} + \frac{5}{12}$

18 $\frac{9}{10} - \frac{1}{4}$

19 $\frac{5}{16} + \frac{3}{12}$

20 $\frac{8}{25} - \frac{1}{10}$

21 $1\frac{3}{8} + 2\frac{1}{4}$

22 $3\frac{1}{2} - 1\frac{2}{3}$

23 $1\frac{1}{3} + 2\frac{3}{5}$

24 $2\frac{3}{4} - 1\frac{5}{8}$

Fractions—Multiplication

Example 1

$\frac{1}{2} \times 6$

same as

$\frac{1}{2} \times 6 = \frac{1 \times 6}{2} = \frac{6}{2}$

$= 3$

Example 3

$1\frac{1}{3} \times 2\frac{1}{4}$

First change to improper fractions

$\frac{4}{3} \times \frac{9}{4}$

$\frac{36}{12} = 3$

1 $1\frac{1}{2} \times 4$ _____

2 $\frac{1}{3} \times 9$ _____

3 $\frac{2}{5} \times 4$ _____

4 $\frac{3}{4} \times 8$ _____

5 $\frac{7}{10} \times 20$ _____

6 $\frac{3}{8} \times 12$ _____

7 $\frac{7}{12} \times 8$ _____

8 $\frac{1}{2} \times \frac{3}{4}$ _____

9 $\frac{5}{8} \times \frac{1}{2}$ _____

10 $\frac{2}{3} \times \frac{1}{4}$ _____

11 $\frac{3}{4} \times \frac{1}{2}$ _____

12 $\frac{1}{3} \times \frac{3}{5}$ _____

13 $\frac{3}{4} \times \frac{8}{3}$ _____

14 $\frac{6}{5} \times \frac{3}{2}$ _____

15 $1\frac{1}{3} \times \frac{3}{4}$ _____

16 $\frac{9}{10} \times \frac{5}{3}$ _____

17 $1\frac{1}{2} \times 1\frac{1}{3}$ _____

18 $\frac{15}{16} \times \frac{2}{5}$ _____

19 $3\frac{1}{4} \times 2\frac{1}{2}$ _____

20 $\frac{15}{16} \times \frac{24}{15}$ _____

Example 2

$\frac{1}{2} \times \frac{4}{5}$

same as

$\frac{1}{2}$ of $\frac{4}{5}$

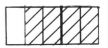

$\frac{1}{2}$ of $\frac{4}{5}$

$=$  $= \frac{2}{5}$

So

Multiply top by top and bottom by bottom

$\frac{1}{2} \times \frac{4}{5} = \frac{1 \times 4}{2 \times 5}$

$= \frac{4}{10} = \frac{2}{5}$

$\frac{2}{3}$ of something is the same as $\frac{2}{3} \times$ that thing

21 Twenty-four people were interviewed. Two-thirds preferred soft spread margarine, but one-quarter did not.
  a How many preferred margarine? _____
  b How many did not? _____
  c How many don't knows? _____

22 A firm employs 96 people. One-third of these are clerical staff, one-eighth managerial, and the rest work in the warehouse.
  a How many are clerical staff? _____
  b How many are managers? _____
  c How many work in the warehouse? _____

Fractions—Division

Example 1

$$4 \div \frac{1}{2}$$

8 halves in
4 wholes

When dividing a
fraction

↓

Turn the fraction
upside down and
multiply

$$4 \div \frac{1}{2} = 4 \times \frac{2}{1}$$

$$= \frac{8}{1} = 8$$

Example 2

$$\frac{3}{4} \div \frac{3}{5}$$

$$\frac{3}{4} \times \frac{5}{3} = \frac{15}{12}$$

$$= \frac{5}{4} = 1\frac{1}{4}$$

1 $3 \div \frac{1}{4}$ _____

3 $8 \div \frac{1}{4}$ _____

5 $27 \div \frac{2}{3}$ _____

7 $\frac{2}{3} \div \frac{1}{3}$ _____

9 $\frac{7}{8} \div \frac{1}{4}$ _____

11 $\frac{5}{8} \div \frac{3}{4}$ _____

13 $\frac{7}{10} \div \frac{3}{5}$ _____

15 $1\frac{1}{2} \div \frac{3}{4}$ _____

17 $2\frac{1}{2} \div \frac{5}{8}$ _____

19 $7\frac{1}{2} \div 1\frac{1}{2}$ _____

2 $2 \div \frac{2}{5}$ _____

4 $11 \div \frac{1}{2}$ _____

6 $\frac{1}{2} \div \frac{1}{4}$ _____

8 $\frac{3}{5} \div \frac{1}{5}$ _____

10 $\frac{4}{5} \div \frac{1}{2}$ _____

12 $\frac{1}{2} \div \frac{2}{3}$ _____

14 $\frac{1}{3} \div \frac{3}{5}$ _____

16 $\frac{7}{8} \div \frac{2}{3}$ _____

18 $\frac{9}{10} \div \frac{3}{4}$ _____

20 $\frac{3}{16} \div \frac{7}{8}$ _____

Example 3

$$2\frac{2}{5} \div 1\frac{1}{4}$$

↓

First change to
top heavy fractions

$$2\frac{2}{5} = \frac{12}{5} \quad 1\frac{1}{4} = \frac{5}{4}$$

$$\frac{12}{5} \div \frac{5}{4}$$

↓

$$\frac{12}{5} \times \frac{4}{5}$$

↓

$$\frac{48}{25}$$

$$1\frac{23}{25}$$

21 A plank 12 m long is to be cut into $\frac{3}{4}$ m
lengths. How many will there be?

_____

22 A ride at a fairground takes $\frac{1}{30}$ of an hour.
How many rides can be given in 3 hours?

_____

23 A pop record programme is on the radio for three-quarters of an
hour. It plays as many new releases as it can. Each record lasts for
one-twentieth of an hour. How many records would you hear on
this programme? _____

Decimals—Addition and Subtraction

Example 1

$$1 \cdot 8 + 10 \cdot 3\,4 + 5 \cdot 3\,7\,9$$

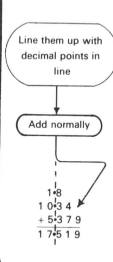

Line them up with decimal points in line

↓

Add normally

```
  1·8
1 0·3 4
+ 5·3 7 9
1 7·5 1 9
```

1
```
  1·3
+ 2·6
```

2
```
  5·7
+ 3·5
```

3
```
1 3·6
−  7·4
```

4 $4\,8 \cdot 4 − 2\,4 \cdot 3 =$ _____

5
```
  5 8·6
+ 2 5·7
```

6 $4\,6 \cdot 2\,5 + 3\,1 \cdot 6\,6 =$ _____

7
```
  3 6·2
− 1 8·9
```

8
```
1 8 9·7
+ 6 3·5
```

9
```
1 0 7·5
−    8·9
```

10 $8\,4 \cdot 3 + 5\,1 \cdot 7 + 6\,2 =$ _____

11
```
  4 6·2 5
− 3 1·6 6
```

12
```
  6 7·3 4
+ 2 9·1 6
```

Example 2

$$2\,3 \cdot 6 − 1\,2 \cdot 8$$

```
  2 3·6
− 1 2·8
  1 0·8
```

Example 3

$$8 \cdot 4 − 1 \cdot 6\,8\,5$$

Put decimal point in

```
8·4 0 0
− 1·6 8 5
  6·7 1 5
```

13
```
1 1 5·0 9
−   2 7·2 5
```

14
```
2 4 6·3 0
  5 5·9 3
+ 2 5·4
```

15 $2\,0\,0 − 5 \cdot 8\,7 =$ _____

16
```
  3 6·2
  2 5·0 0
1 4 3·0 5
+   6·8
```

17
```
1 7 6·5
−  2 7·9 5
```

18
```
  7 6 1·3 2
−  5 9 9·6 8
```

19
```
3 1 2·6 5
−   2 8·4 6
```

20 $3\,2\,6 \cdot 1 + 1\,0 \cdot 9 + 1\,7 \cdot 6\,8 + 1\,9 \cdot 3\,9 =$ _____

21 In a supermarket I buy articles costing:
£1·73, 53p, £2·25, £0·37, and £4·10
  a How much do these cost altogether? _____
  b How much change from £10? _____

22 The cost of a bag of sugar in different shops is:
33p, £0·32, 35p, £0·31 and 34p.
What is the average cost of a bag of sugar? _____

14

Decimals—Multiplication

Decimal numbers are multiplied by 10, 100, 1000 . . . by moving
the decimal point to the **right** by 1, 2, 3 . . . places.

Multiply by 10

$4 \cdot 5 \times 10 = 4 \cdot 5 = 4 5$

$1 \cdot 8 9 \times 10 = 1 \cdot 8 9 = 1 8 \cdot 9$

Multiply by 100

$4 \cdot 5 \times 100 = 4 \cdot 5 0 = 4 5 0$

$1 \cdot 8 9 \times 100 = 1 \cdot 8 9 = 1 8 9$

Example 1

$1 5 \cdot 9 \times 7$

(Multiply as normal)

$$\begin{array}{r} 1 5 \cdot 9 \\ \times 7 \\ \hline 1 1 1 3 \end{array}$$

(Where does the decimal point go?)

(Count figures to the right of the decimal point)

(Count back the same number in the answer)

There is only 1 figure
($\cdot 9$) so count back
one

($1 1 1 \cdot 3$)

The answer is $1 1 1 \cdot 3$

---

1 Complete these tables:

a

3·8	×10	38
2 6·8		
1 3·2		
1 5		
1·6 5		
1 2 8·9		
1 5·6 2		
0·0 6 5		

b

3·8	×100	380
2 6·8		
1 3·2		
1 5·		
1·6 5		
1 2 8·9		
1 5·6 2		
0·0 6 5		

Example 2

$1 2 6 \cdot 3 5 \times 2 \cdot 4$

(Multiply as normal)

$$\begin{array}{r} 1 2 6 3 5 \\ \times 2 4 \\ \hline 5 0 5 4 0 \\ 2 5 2 7 0 0 \\ \hline 3 0 3 2 4 0 \end{array}$$

There are 3 figures
on the right of the
decimal point
($\cdot 3 5$ and $\cdot 4$)

(Count back 3)

($3 0 3 \cdot 2 4 0$)

The answer is

$3 0 3 \cdot 2 4 0$

2
$$\begin{array}{r} 7 \cdot 3 \\ \times 2 \\ \hline \end{array}$$

3
$$\begin{array}{r} 6 \cdot 5 \\ \times 6 \\ \hline \end{array}$$

4
$$\begin{array}{r} 1 4 \cdot 8 \\ \times 5 \\ \hline \end{array}$$

5
$$\begin{array}{r} 3 7 \cdot 2 \\ \times 8 \\ \hline \end{array}$$

6
$$\begin{array}{r} 7 2 \cdot 6 \\ \times 9 \\ \hline \end{array}$$

7
$$\begin{array}{r} 1 5 3 \cdot 6 \\ \times 5 \\ \hline \end{array}$$

8
$$\begin{array}{r} 6 3 1 \cdot 8 \\ \times 1 3 \\ \hline \end{array}$$

9
$$\begin{array}{r} 1 2 5 \cdot 6 \\ \times 2 9 \\ \hline \end{array}$$

10
$$\begin{array}{r} 5 4 \cdot 9 \\ \times 3 \cdot 5 \\ \hline \end{array}$$

11
$$\begin{array}{r} 1 2 \cdot 9 \\ \times 4 \cdot 4 \\ \hline \end{array}$$

12
$$\begin{array}{r} 4 1 9 \cdot 8 \\ \times 7 \cdot 3 \\ \hline \end{array}$$

13
$$\begin{array}{r} 2 3 5 \cdot 1 \\ \times 1 2 \cdot 6 \\ \hline \end{array}$$

14
$$\begin{array}{r} 6 0 8 \cdot 7 \\ \times 2 3 \cdot 4 \\ \hline \end{array}$$

15
$$\begin{array}{r} 5 6 8 \cdot 4 \\ \times 7 \cdot 8 \\ \hline \end{array}$$

16
$$\begin{array}{r} 8 3 \cdot 2 6 \\ \times 8 \cdot 1 \\ \hline \end{array}$$

17
$$\begin{array}{r} 5 0 \cdot 0 3 \\ \times 0 \cdot 6 \\ \hline \end{array}$$

18
$$\begin{array}{r} 3 9 \cdot 6 0 2 \\ \times 4 \cdot 9 \\ \hline \end{array}$$

19
$$\begin{array}{r} 3 0 2 \cdot 5 6 \\ \times 1 7 \cdot 8 \\ \hline \end{array}$$

20 In a sale one pair of socks is marked at £1·62 and ten pairs are offered
at £15·35. How much do you save if you buy the ten pairs?

Decimals—Division

Decimal numbers are divided by 10, 100, 1000 . . . by moving
the decimal point to the **left** by 1, 2, 3 . . . places.

Divide by 10

$1\ 5\ 5 \cdot 3 \div 1\ 0 = 1\ 5\ 5 \cdot 3 = 1\ 5 \cdot 5\ 3$

$2 \cdot 6\ 7 \div 1\ 0 = 2 \cdot 6\ 7 = 0 \cdot 2\ 6\ 7$

Divide by 100

$1\ 5\ 5 \cdot 3 \div 1\ 0\ 0 = 1\ 5\ 5 \cdot 3 = 1 \cdot 5\ 5\ 3$

$2 \cdot 6\ 7 \div 1\ 0\ 0 = 0\ 2 \cdot 6\ 7 = 0 \cdot 0\ 2\ 6\ 7$

Example 1

$1\ 6 \div 5$

(Write it like this)

$5\,\overline{)1\ 6}$

(Put in decimal points and some noughts)

$5\,\overline{)1\ 6 \cdot 0\ 0}$

(Divide as normal)

$\begin{array}{r} 3 \cdot 2 \\ 5\,\overline{)1\ 6 \cdot 0\ 0} \end{array}$

1 Complete these tables:

a

2 4 7·8	÷ 1 0	2 4·7 8
5 6·1		
2 5·2 5		
1 7·6 2		
8·9 2		
1 4 0·4		
1·2 8 9		
6·0 1 8		

b

2 4 7·8	÷ 1 0 0	2·4 7 8
5 6·1		
2 5·2 5		
1 7·6 2		
8·9 2		
1 4 0·4		
1·2 8 9		
6·0 1 8		

2 $1\ 4 \div 4$

3 $3\ 7 \div 5$

4 $1\ 0\ 0 \div 8$

5 $9\ 2 \cdot 4 \div 7$

6 $4\ 6\ 7 \cdot 2 \div 8$

7 $2\ 8 \cdot 6\ 1\ 5 \div 5$

8 $2\ 4\ 7 \cdot 2\ 7 \div 4$

9 $2 \cdot 6\ 5\ 3\ 7 \div 7$

10 $4\ 6 \cdot 9\ 8 \div 0 \cdot 9$

11 $1\ 5 \cdot 6 \div 1 \cdot 2$

12 $2\ 6 \cdot 0\ 1 \div 5 \cdot 1$

13 $7\ 6\ 8 \cdot 4\ 2 \div 0 \cdot 6$

Example 2

$4 \cdot 3\ 2 \div 1 \cdot 6$

$\dfrac{4 \cdot 3\ 2}{1 \cdot 6}$

(We want a whole number on the bottom)

(Multiply top and bottom by 1 0)

$\left(\dfrac{4 \cdot 3\ 2}{1 \cdot 6} = \dfrac{4\ 3 \cdot 2}{1\ 6} \right)$

$\begin{array}{r} 2 \cdot 7 \\ 1\ 6\,\overline{)4\ 3 \cdot 2} \end{array}$

Example 3

$1\ 1 \cdot 4\ 8 \div 0 \cdot 8$

$\left(\dfrac{1\ 1 \cdot 4\ 8}{0 \cdot 8} = \dfrac{1\ 1\ 4 \cdot 8}{8} \right)$

$\begin{array}{r} 1\ 4 \cdot 3\ 5 \\ 8\,\overline{)1\ 1\ 4 \cdot 8\ 0\ 0} \end{array}$

14 Six people all contribute the same amount to
a football pools coupon. The total stake is
£2·40, and one week they win £31·20.

a How much do they each contribute?

_____

b How much do they each win?

_____

c How many times greater than the stake
are the winnings? _____

Decimals—Conversion

A decimal is a fraction with a denominator of 10, 100, 1000 . . .

0·7   is the same as $\frac{7}{10}$

0·0 7   is the same as $\frac{7}{100}$

0·0 0 7 is the same as $\frac{7}{1000}$

Example 1

Fractions to decimals

$\frac{2}{5}$

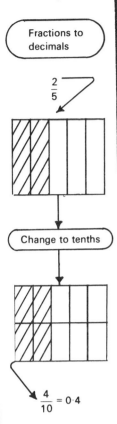

Change to tenths

$\frac{4}{10} = 0·4$

Example 2

$\frac{3}{8} = 3 \div 8$

$$\begin{array}{r} 0·3\,7\,5 \\ 8\overline{)3·0\,0\,0} \end{array}$$

Example 3

Decimals to fractions

$0·6 = \frac{6}{10}$

$\frac{6}{10} = \frac{3}{5}$

Example 4

0·4 2 5

$= \frac{4}{10} + \frac{2}{100} + \frac{5}{1000}$

$= \frac{400}{1000} + \frac{20}{1000} + \frac{5}{1000}$

$\frac{425}{1000} = \frac{85}{200}$

$\frac{85}{200} = \frac{17}{40}$

1 Complete the table below:

Fraction	Working	Decimal
$\frac{5}{8}$		
$\frac{5}{6}$		
$\frac{1}{2}$		
		0·5 5
		0·1 8
$\frac{3}{5}$		
		0·3
		0·1 2 5
		0·1 5
$\frac{7}{10}$		
$\frac{3}{4}$		
$\frac{8}{20}$		
$\frac{2}{7}$		
		0·3 5
		0·5 2 5
$\frac{7}{25}$		
		0·2 5

Sometimes when changing a fraction to a decimal the division does not stop:

$$\frac{1}{3} = 1 \div 3 \text{ so } 3\overline{)1·0\,0\,0\,0\,0\,0}^{0·3\,3\,3\,3\,3\,3} \cdots$$

Know when to stop! Eg £2·6 6 6 . . . would be given as £2·6 7

2 How many pence is closest to two-thirds of £1 ? _____

3 You are asked for $\frac{5}{6}$ of £1000. What would you give ? _____

4 An insurance company pays out only seven-twelfths of any claim. You make a claim for £2,500. How much would you expect to receive ? _____

Decimals—Percentages

A percentage is a fraction with **a denominator of 100**.

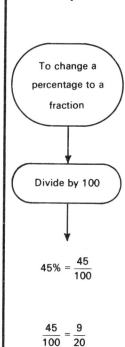

Example 1

$60\% = \dfrac{60}{100}$

$\dfrac{60}{100} = \dfrac{6}{10} = \dfrac{3}{5}$

Example 2

To change a percentage to a fraction

↓

Divide by 100

↓

$45\% = \dfrac{45}{100}$

$\dfrac{45}{100} = \dfrac{9}{20}$

1 Convert these percentages to fractions in lowest terms.

a 80% =

b 30% =

c 35% =

d 40% =

e 50% =

f 70% =

g 85% =

h 22% =

i 46% =

j 72% =

Example 3

To change a fraction to a percentage

↓

multiply by 100

↓

$\dfrac{2}{5} = \dfrac{2}{5} \times 100\%$

$= \dfrac{200}{5}\%$

$= 40\%$

2 Convert these fractions to percentages.

a $\dfrac{1}{5} =$

b $\dfrac{3}{5} =$

c $\dfrac{1}{4} =$

d $\dfrac{3}{4} =$

e $\dfrac{7}{10} =$

f $\dfrac{3}{8} =$

g $\dfrac{13}{20} =$

h $\dfrac{17}{40} =$

i $\dfrac{19}{50} =$

j $\dfrac{1}{3} =$

k $\dfrac{2}{3} =$

l $\dfrac{7}{8} =$

3 Approximately one-quarter of the cars examined in a survey are Minis. What percentage is this? _____

4 35% of goods manufactured in Britain are exported. What fraction of our output is this? _____

5 Of every 1000 babies born, 520 are boys. What percentage are girls? _____

6 Two-fifths of the adult population of Britain need glasses. What percentage is this? _____

Decimals—Percentages of

Example 1

60% of £20?

Express % as a fraction

$60\% = \dfrac{60}{100} = \dfrac{3}{5}$

$\dfrac{3}{5}$ of £20

$= \dfrac{3}{5} \times 20 = \dfrac{60}{5}$

$= £12$

1 Find the following:

a 50% of £10 _____

b 20% of 200 g _____

c 35% of 400 kg _____

d 60% of 120 m _____

e 70% of 80 minutes _____

f 95% of 500 miles _____

g 44% of 250 volts _____

h 36% of 2 500 km _____

i $22\frac{1}{2}\%$ of £200 _____

j $66\frac{2}{3}\%$ of £210 _____

Example 2

In a survey of 600 cars, 150 were red. What % is this?

Form fractions

$\dfrac{150}{600} = \dfrac{15}{60} = \dfrac{3}{12}$

$= \dfrac{1}{4}$

$\dfrac{1}{4} \times 100\% = \dfrac{100}{4}\%$

$= 25\%$

2 A class did a survey on local traffic. Altogether they counted 240 cars. Of these, 50% had only the driver in, and 160 were heading East.
  a How many were heading West? _____
  b What percentage were heading West? _____
  c How many drivers were carrying passengers? _____

3 In a school of 860 pupils, 55% are girls.
  a How many girls are there? _____
  b How many boys are there? _____
  c The pupils are split equally into five houses.
     What percentage is in each house? _____
  d There are 150 pupils in the fifth form.
     What percentage of the school is this? _____
     (to the nearest whole %)

4 A factory produces 'Stereo Centres' which it sells for £250. It reckons that 35% of this goes in wages, 40% in materials, 10% in other costs, and the rest is profit.
  a How much do the materials cost for one Stereo Centre? _____
  b What percentage is profit? _____
  c How much profit do they make on ten Stereo Centres? _____

19

Decimals—Approximation

Sometimes, to make calculations easier to handle, precise figures can be approximated to more convenient figures. For example:

  5 cm 3 mm can be approximated to 5 cm

  3·2 volts can be approximated to 3 volts

  A 24-h time of 12 21 39 can be approximated to 12.22

Careful judgement is required when approximating figures.

1 Approximate the following measurements to the accuracy required:

a 7·9 m to the nearest m _____   **f** 27·6 cm to the nearest cm _____

b 55·2 g to the nearest g _____   **g** 4·55 mm to the nearest mm _____

c 2166 cm to the nearest m _____   **h** 458 cm to the nearest m _____

d 50 inches to the nearest foot _____   **i** 12.45 to the nearest hour _____

e 158 732 to the nearest thousand _____   **j** Five fifty three to the nearest half hour _____

2 Measure these bars and give the answer to the nearest   **a** cm   **b** inch.

a _____   **b** _____

a _____   **b** _____

3 What is the reading on this meter to the nearest quarter unit? _____

4 Give two different times this clock could be indicating to the nearest minute.

_____   _____

Decimals—Estimation

In finding an estimate to a problem, each part of it is replaced by a more convenient approximation. When estimating in this way, we use the approximate sign $\simeq$ rather than the equal sign $=$.

Example 1

19×11

Approximately

$19 \simeq 20$
$11 \simeq 10$

So

$19 \times 11 \simeq 20 \times 10$
$\simeq 200$

Example 2

$\dfrac{5 \cdot 1 \times 19 \cdot 2}{9 \cdot 8}$

$\simeq \dfrac{5 \times 20}{10}$

$\simeq \dfrac{100}{10}$

$\simeq 10$

1 Give approximate answers to these:

a $9 \times 11 \simeq$ _____

b $18 \times 32 \simeq$ _____

c $105 \times 11 \simeq$ _____

d $349 \times 9 \simeq$ _____

e $9 \cdot 3 \times 21 \simeq$ _____

f $548 \times 1 \cdot 9 \simeq$ _____

g $128 \cdot 5 \times 4 \cdot 8 \simeq$ _____

h $103 \times 9 \simeq$ _____

i $29 \div 4 \cdot 9 \simeq$ _____

j $11 \cdot 2 \times 28 \cdot 7 \simeq$ _____

k $\dfrac{11 \times 19}{9} \simeq$ _____

l $\dfrac{197 \times 9 \cdot 2}{11} \simeq$ _____

m $\dfrac{58 \cdot 9 \times 1 \cdot 9}{5 \cdot 2} \simeq$ _____

n $\dfrac{21 \cdot 2 \times 102}{18 \cdot 9 \times 5 \cdot 1} \simeq$ _____

2 Underline the approximate amount needed to buy the following.

    a 12 tins of beans at $11\frac{1}{2}$p each. £1·00   £1·50   £2·00
    b 8 books at £1·23 each. £8·00   £10·00   £12·00
    c 1 bottle of lemonade if 6 cost £1·74. 20p   30p   40p

3 Rolls of wallpaper are 50 cm wide and 11 m long. The average room is 5 m × 3·2 m and 3 m in height.
    a How many strips of ceiling height can be cut from a roll? _____
    b How many rolls are needed to paper the average room? _____

Money—Basics

1 Fill in the table below to show how you could pay for each item.

Item	Cost	£5	£1	50p	10p	5p	2p	1p	
Newspaper	10p				√				These are
Coat	£24·99	√ √ √ √	√ √ √ √	√	√ √ √ √	√	√ √		done
Toothbrush	23p								for you.
Diary	75p								
Hat	£4·24								
Sweater	£15·15								
LP	£4·16								
Disco	85p								
Film	£1·20								
Outing	£7·37								
Present	£2·57								
Shoes	£11·56								
Football match	£1·25								
Portable radio	£29·35								

Example 1

£1·82 + £23·15 + £0·13

(Line them up first)

```
  £1·8 2
  2 3·1 5
+ 0·1 3
─────────
 £25·1 0
```

2
```
  £7 2·6 3
+ £3 6·4 4
───────────
```

3
```
  £1 5·6 5
+ £8·7 1
───────────
```

4 £7 6 + £3 4·0 1 + £2 0·0 9 _____

5 £1 2 3 − £4 9·5 4 + £2·2 7 _____

6
```
  £5·6 1
    × 6
─────────
```

7 £1 2 3·7 2 × 1 2

8 £6·4 4 ÷ 7

9 £6 3 2·5 1 × 2 7

Example 2

£4 3·5 6 × 7

```
  £4 3·5 6
       × 7
───────────
 £3 0 4·9 2
```

Example 3

£2 6 8·5 6 ÷ 1 2

```
       2 2·3 8
 1 2)2 6 8·5 6
```

£2 2·3 8

Money—Shopping I

SAVINGS ON FRUIT AND VEGETABLES

28 lb Selected Potatoes	28 lb bag	**£1·15**
English Conference Pears	2 lb bag	**28p**
English Cox Apples	3 lb bag	**39p**

1 a How much are pears per pound? _____ **b** How much are apples per pound? _____

   c How much are potatoes per pound (to the nearest penny)? _____

   d Find the cost of each item   2 28 lb bags of potatoes _____

      and the total on this   6 lb of pears   _____

      shopping list.   6 lb of apples   _____

                                                 Total _____

2 My weekly paper bill is £1·32 but I pay every 4 weeks. How much do I pay? _____

3 If I buy 4 records at £1·15 each, how much change do I get from £5? _____

4 a How many packets of tea at 24p each can I buy for £2? _____

   b How much is left over? _____

5 I take a girl to the pictures. The bus fare is 45p (single), and it costs £1·25 each to get in. If I pay for both of us, how much will the evening out cost me? _____

6 When I went out I had £7·32$\frac{1}{2}$, when I got home I only had £1·58$\frac{1}{2}$ left.

   How much did I spend? _____

7   COZY PIPE INSULATION

For 15 mm copper pipe, in 1 metre lengths at	**36p each**
For 22 mm copper pipe, in 1 metre lengths at	**44p each**
For 28 mm copper pipe, in 1 metre lengths at	**57p each**
Frostguard	**68p a roll**
Micafil	**£2.35 a bag**

   Find how much it will cost if I need the following items to insulate my bathroom:

   a 2 metres of pipe insulation for 15 mm pipe _____

   b 3 metres of pipe insulation for 22 mm pipe _____

   c 1 roll of Frostguard   _____

   d 1 bag of Micafil   _____

   e The total cost is   _____

8 Find the sale price of the items in the table.

	Original price	Sale price
Colour T.V.	£340	
Bicycle	£80	
Fur coat	£720	

Money—Shopping II

1 Advertisement Rates

| £1 for the first 8 words and 10p for each subsequent word. |

Find the cost of these advertisements.

a Triumph Bonneville 1976 Good condition. New tyres. Rack, Top Box. Bargain £650. Tel 01-752 5254. Cost is _____

b Honda Goldwing Executive. Immaculate condition. Only 1500 miles. Many extras. House purchase forces sale. £2200. No offers. Box 4080. Cost is _____

2 A Chinese Meal

Sheila is sent to the Take-Away to fetch the family supper.

The Golden Dragon

6	*Golden Dragon Chop Suey*	£1·20
13	*Sweet and Sour King Prawn*	£1·05
16	*Curried Beef*	£0·95
22	*Golden Dragon Chow Mein*	£1·15
32	*King Prawn Egg Foo Yung*	£1·00
36	*Chips*	£0·25
	Meal for 4 persons	£5·85

a Find the cost of each person's meal if this is what is ordered.

Dad: 6, 32, 36 _____

Mum: 16, 36 _____

David: 13, 36 _____

Sheila: 22, 32 _____

b What is the total bill? _____

c How much change do they get from £10? _____

d How much cheaper would it be to have the meal for 4? _____

3 T.V. Rental   (Note: ″ is the symbol for inches)

	Annual plan	Monthly plan
Screen size	**Annual payment**	**Monthly payment**
19/20″ colour	£94·50	£8·75
26″ colour	£119·56	£11·07
20″ monochrome	£42·12	£3·90

a On the annual plan you must pay half the payment as a deposit. How much is this for

19/20″ colour? _____

26″ colour? _____

20″ monochrome? _____

b On the monthly plan how much would you pay for each T.V. a year?

19/20″ colour? _____

26″ colour? _____

20″ monochrome? _____

c How much more is this than on the annual plan in each case?

19/20″ colour? _____

26″ colour? _____

20″ monochrome? _____

d Divide the annual plan prices by 52 to find the equivalent weekly cost in each case. Give your answers to the nearest penny. Multiply the monthly plan prices by 12, then divide by 52, to find the equivalent weekly cost in each case.

	Annual plan Weekly cost	Monthly plan Weekly cost
19/20″ colour		
26″ colour		
20″ monochrome		

Money—Hire Purchase

Hire Purchase (HP). This means paying a deposit and then paying the rest (called the **Balance**) in monthly instalments.

GVM 4554 MUSIC CENTRE

A Dolby music centre for under £180, 12 watts, r.m.s. 3 band radio. Speakers included.

OUR PRICE
£179.95
or £18.00 deposit and 9 monthly payments of £20.32.

1 This music centre can be bought for either £179·95 cash, or for an £18·00 deposit and nine payments of £20·32.

   a What is the total HP price?

_____

   b How much more expensive is the HP price than the cash price?

_____

2 a What is the cash price of this washing machine?

_____

   b What is the HP price?

_____

   c How much cheaper would it be to pay cash?

_____

MOTION 60

Stainless steel tub and drum. Quiet induction motors. Zinc coated, rustproofed cabinet – proven reliability. Lots more quality for little more money.
MRP £335.00
OUR PRICE
£289.00
Deposit £57.80
24 monthly payments of £13.00.

MM BIKES

	Price	Deposit	Months	Per Month
Mercury	£2995	£450	36	£86.99
Nomad	£2095	£315	36	£60.70

The greater the deposit – the lower the repayments.

3 a What is the HP price of the Nomad?

_____

   b In a sale, these prices are reduced by 15%. What is the new cash price of the Nomad?

_____

   c What is the HP price of the Mercury?

_____

   d If you paid a deposit of £1000 for the Mercury and you still paid the same HP price, what would be the new monthly repayment?

_____

Money—Bank Account

1 Here is a bank statement for Albert Stubbs.

Statement				CREDIT BANK LTD
A. Stubbs				The Square Preston
A/C no. 8654341				
date 22.12.81				

Date	Description	Debits	Credits	Balance
Nov 22				135·60
Nov 23	wages		323·85	
Nov 25	087646	23·00		
Nov 29	Jones and Co	16·80		
Dec 2	087647	50·00		
Dec 5	087649	2·50		
Dec 9	087648	50·00		
Dec 10	087650	185·72		
Dec 11	087651	13·27		
Dec 13	087653	50·00		
Dec 15	087652	8·62		
Dec 20	087654	50·00		
Dec 21	Lancs C.C.		15·16	

a How much was in the account on Nov 25? _____

b Fill in the balance after each credit and debit until Dec 21.
How much is left on that date? _____

c Mr Stubbs writes this cheque on Dec 22.

CREDIT BANK LTD	Dec 22 - 1981
Pay J. Dean & Co Ltd	
Amount Thirty Five Pounds	£ 35 - 00
8654341–0130090–087655	a. Stubbs.

How much is left in the account when this cheque is drawn? _____

d On Dec 23 his wages are paid in. If he is paid the same amount as last month, what balance does this leave in the account? _____

e Albert then writes the following cheques: £52·60 on Dec 28; £150·26 on Jan 1; £35·14 on Jan 8 and £14·26 on Jan 9. He receives a cheque for £32·16 on Jan 10. Complete his statement below.

Date	Description	Debits	Credits	Balance
Dec 22				
Dec 23				
Dec 28				

Money—Counters

1 In a plastics factory, a counter records how many units (finished articles) the machine makes in an hour.

At the beginning of the day the reading is `1 2 4 5 0`

The readings at the end of every hour are

`1 3 7 0 0` `1 5 0 5 0` `1 6 4 8 0` `1 7 7 5 3`

a Complete the table to show how many units were produced in each hour and the total for the whole morning.

1st hour	2nd hour
13700 −12450 ————	————
No of units	No of units
3rd hour	**4th hour**
————	————
No of units	No of units

Total = _____ units

b Each unit is sold for 40p.
How much money does the machine make in the morning? _____
c The machine makes the same number in the afternoon.
How many will it make in a day? _____
d The same number are made every day.
How many will be made in a 5-day week? _____
e How much money will the machine make in a week? _____

2 Gas consumption is recorded by the dials on the gas meter. Some gas meters look like this one, which shows a reading of 1500 therms (units of gas).

a Complete the drawing below to show a reading of 1610.

b Gas meters are usually read every quarter. If the reading was 1500 at the beginning of a quarter and 1610 at the end, how many therms have been used? _____
c Each quarter there is a standing charge of £1·62. The first 52 therms cost 20·8p each and further therms cost 16·5p each. Use this information to complete the gas bill below.

Standing charge £   1·62
52 therms at 20·8p each £ _____
__ therms at 16·5p each £ _____
Total charge £ _____

Money—Wages I

Spanner and Co. are an engineering firm.
Stan Smith and Fred Riley are fitters. Their rates of pay are;

   Basic: £2·20 per hour for 40 hours.
Overtime: First two hours at time and a quarter.
         Next two hours at time and a half.
         Any other at double time.

Each employee is allowed 3 minutes clocking-in time per day. No deductions from pay are made for this time.

a Here are their clock cards for one week. Complete them.

Stan

	AM		PM		
	IN	OUT	IN	OUT	Hours
Mon	08.00	12.00	13.00	17.00	8
Tues	08.02	12.00	13.01	17.00	
Wed	07.59	12.00	13.00	17.00	
Thurs	08.00	12.00	13.01	17.00	
Fri	08.00	12.00	13.00	17.30	
Sat	08.00	11.30			
Sun			14.00	16.00	
			Total	Hours	

Fred

	AM		PM		
	IN	OUT	IN	OUT	Hours
Mon	08.01	12.00	13.00	17.30	$8\frac{1}{2}$
Tues	08.00	12.00	13.01	17.30	
Wed	08.00	12.00	13.00	17.00	
Thurs	07.59	12.00	12.59	17.00	
Fri	07.58	12.00	13.00	17.00	
Sat	08.00	13.00			
Sun	08.00	12.00			
			Total	Hours	

b Complete these tables and work out their wages for the week.

Stan		Fred	
40 hours @ £2·20	= _____	40 hours @ £2·20	= _____
hours @ $1\frac{1}{4}$ (£2·75)	= _____	hours @ $1\frac{1}{4}$ (£)	= _____
hours @ $1\frac{1}{2}$ (£)	= _____	hours @ $1\frac{1}{2}$ (£)	= _____
hours @ 2 (£)	= _____	hours @ 2 (£)	= _____
Total	= _____	Total	= _____

c Stan's payslip is shown below. He pays tax at the rate of 25% after his N.I. and Pension contributions have been deducted. Fill in the missing entries.

```
              SPANNER AND CO.
Name: Stan Smith               Number: 0040096
                     N.I. No.: YL/44/19/77B

Tax Code: 116L*
Gross Earnings £
   N.I.        £ 8·00        Check that your
   Pension    £12·35         pay packet contains
   Tax                       this amount.
              Net payment  £ _____
```

Money—Wages II

1 Mrs Hardaker makes dresses, skirts and blouses in a textile factory.
She normally works 40 hours each week.
A dress takes 30 minutes to make and earns her 75p.
A skirt takes 15 minutes to make and earns her 30p.
A blouse takes 45 minutes to make and earns her £1·20.

a One week she makes 40 dresses,
47 skirts and 11 blouses.
Complete this week's work card.

Item	No.	Time	Earnings
Dresses	40		
Skirts	47		
Blouses	11		
	Total		

b The next week she spends $11\frac{1}{2}$
hours on dresses, 15 hours on
skirts, and $13\frac{1}{2}$ hours on blouses.
Complete this week's work card.

Item	No.	Time	Earnings
Dresses		$11\frac{1}{2}$ hr	
Skirts		15 hr	
Blouses		$13\frac{1}{2}$ hr	
	Total		

c The week following, there is only
one order for 280 dresses and this
is equally divided between 7
women. When there is not
enough work to do, they each
get 50p per hour waiting time.
Complete this week's work card.

Item	No.	Time	Earnings
Dresses			
Skirts			
Blouses			
Waiting		—	
	Total		

2 In the packing section of a biscuit factory the employees are paid
at a basic rate of £54 for packing 1600 boxes with 15
packets per box. For every box over 1600 they are paid a
'piece work' bonus of 1p.
a How many packets do they
box at the basic rate? _____
b How much are they paid
for packing each packet at the basic rate? _____
c How many packets must be
packed to earn a wage of £60? _____

Money—Costing

Smith Engineering make boilers for dying fabrics. The cost of making a boiler is found by adding the labour costs and the material costs.

Labour costs

Each boiler involves the labour of a fitter (F), an electrician (E), a welder (W) and an inspector (I). Below is the work card for one week's work on a boiler.

Time	8–9	9–10	10–11	11–12		1–2	2–3	3–4	4–5
Mon	F	F	W	W		F	W	F	I
Tues	W	F	E	F		F	F	I	W
Wed	W	F	E	W		F	F/E	F	W
Thurs	E	E	F	W		I	F	F	W
Fri	E	F	W	W		E	F	F	I

F/E means $\frac{1}{2}$ hour each.

a Complete this table to find the labour costs for one week.

	Total hours worked	Rates of pay per hour	Total cost
Fitter		£1·75	hrs @ £1·75 =
Welder		£1·85	hrs @ £1·85 =
Electrician		£2·05	hrs @ £2·05 =
Inspector		£2·30	hrs @ £2·30 =
			Total

b A boiler takes six weeks to complete.
What is the total labour cost? _____

Material costs

c Complete this table to find the material costs.

Material	Cost	Total
4 sheets of 2 m × 1·5 m aluminium sheet	£1·60 per m²	
85 m electric wire	£20 per 30 m roll	
7 valves	£14·75 each	
150 bolts	£15·00 per 100	
75 washers	£10·00 per 1000	
75 litres of acetylene gas for welding	£2·50 per litre	
	Total	

d Find the total cost of labour and materials.

e Add on 10% for overheads

f Add on 10% for profit

g The selling price is £_____

labour £
materials £_____
total £
+ 10% _____
_____
+ 10% _____
total _____

30

Weights and Measures

Facts

Metric units

10 mm = 1 cm

100 cm = 1 m

1000 m = 1 km

1000 g = 1 kg

Approximately

10 cm ≃ 4 in

1 m ≃ 40 in

100 km ≃ 60 mi

1 kg ≃ 2·2 lb

Facts

Imperial units

12 inches = 1 foot (ft)
(in)

3 ft = 1 yard (yd)

1760 yd = 1 mile (mi)

16 ounces = 1 pound
(oz)   (lb)

Approximately

1 ft ≃ 30 cm

100 mi ≃ 160 km

1 lb ≃ 450 g

Use the information in the tables to answer the questions.

1 28 in = _____ ft _____ in

2 55 in = _____ ft _____ in

3 How many feet in 1 mile? _____

4 37 oz = _____ lb _____ oz

5 66 mm = _____ cm

6 535 cm = _____ m

7 34 mm = _____ cm   **8** 187 mm = _____ cm

9 468 cm = _____ m   **10** 763 cm = _____ m

11 1255 cm = _____ m   **12** 5 m = _____ cm

13 4000 g = _____ kg   **14** 2107 g = _____ kg   **15** 7·82 kg = _____ g

In the following questions, give approximate answers to the nearest whole unit.

16 5 in ≃ _____ cm   **17** 9 cm ≃ _____ in   **18** $1\frac{1}{2}$ in ≃ _____ cm

19 7 cm ≃ _____ in   **20** $2\frac{1}{2}$ lb ≃ _____ kg   **21** 1 m ≃ _____ ft

22 A man is exactly 6 feet tall. This is roughly _____ cm or _____ m.

23 A football pitch is 42 m wide. This is _____ ft.

24 From Derby to Glasgow is about 250 miles. This is _____ km.

25 How many grams does half a pound of coffee weigh? _____ g.

26 A room is 4·5 metres wide. What is this in _____ ft and _____ in?

27 Athletes run a 100 m race. This is _____ yards _____ ft _____ in.

28 One of the stages in the 'Tour de France' is 150 km. This is _____ mi.

29 The average man weighs 85 kg. This is _____ lb.

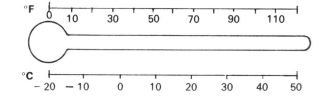

30 Use a ruler to answer these questions.
  a 0° Celsius = _____ ° Fahrenheit
  b 50°F = _____ °C
  c Room temperature is about 21 °C.
     This is about _____ °F.

31

Bar Charts

1 This bar chart shows the goals scored by the teams in the first and second football divisions in one Saturday.

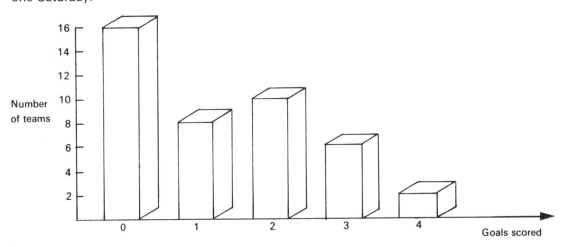

  a Which was the most common score? _____

  b How many teams scored that number of goals? _____

  c How many teams scored two goals? _____

  d How many goals were scored on this Saturday? _____

  e How many teams played altogether? _____

  f What was the average number of goals per team? _____

2 Here is a Double bar chart. It shows the sales (in thousands) of two cars, the BIRD and the SPORT, from January 1975 to December 1978.

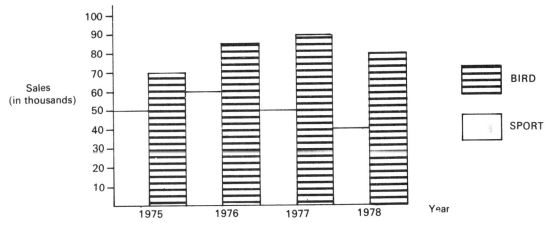

  a How many people bought a BIRD in 1976? _____

  b How many people bought a SPORT in 1978? _____

  c How many more BIRDS than SPORTS were sold in 1975? _____

  d How many fewer people bought SPORTS than BIRDS in 1977? _____

  e How many BIRDS were sold in these years? _____

  f How many SPORTS were sold in these years? _____

  g What were the average sales of the BIRD per year? _____

  h What were the average sales of the SPORT per year? _____

Food Arithmetic

A balanced diet is one which has an adequate energy value and contains all the nutrients the body needs in the required proportions. The energy value of food is measured in calories. The table below gives approximately the number of calories in a variety of foods.

Food	Calories per 100 g
Stew	250
Meat	300
Chicken Supreme	225
Fish	200
Green Vegetable	40
Mash	80
Chips	250
Roast Potato	120
Meat Pie	300

Food	Calories
Puddings	250 per portion
Ice Cream	100 per portion
Eggs	80 each
Cornflakes	70 per portion
Mars Bar	270 each
Sandwiches	90 each
Beer	200 per pint
Chocolate Eclair	200 each
Cheese and Biscuits	150 per portion
Coffee/Tea (white)	15 per cup
Sugar	25 per teaspoon

1 A works canteen has about 80 workers to stay for lunch. Here is the menu for one week.

Mon	Meat pie, vegetables, roast potatoes, apple pie
Tues	Stew, vegetables, mashed potato, chocolate eclair
Wed	Pork, vegetables, mashed potato, bread and butter pudding
Thurs	Chicken supreme, vegetables, sponge pudding
Fri	Fish, vegetables, chips, cheese and biscuits

Each meal has about 100 g of meat or fish, 100 g of vegetables and 150 g of potato.
a Work out how many calories in Monday's meal.

   meat _____ vegetables _____ potatoes _____ sweet _____
b If the average adult man requires 2500 calories per day, what fraction of his calorie requirement would he get from Monday's meal? _____
c How much fish should the canteen order for Friday? _____
d Roughly, how much potato does the canteen serve in a week **(i)** in grams? _____
(ii) in kilograms? _____ **(iii)** in pounds? _____
e If the canteen uses three types of vegetables, how much of each should be ordered for a 5-day week? _____

2 An average man (who needs 2500 calories per day) consumes the following in the course of one day. Fill in how many calories each of these contain.

Breakfast—cornflakes _____   Lunch—4 sandwiches _____
Dinner—300 g stew, 200 g chips _____
Extras—6 cups of tea with milk and 2 sugar, 1 Mars bar and 2 pints of beer _____
Is he over- or under-eating, and by how much? _____

Clocks and Time

On the 24 hour clock, the times are the same until midday (12.00 hours). Then we count on until 24.00 hours (midnight).

1 Convert these times to 24 hour clock times.

   a 6.30 am _____ **b** 4.00 pm _____ **c** 11.46 am _____ **d** 7.58 pm _____

   e Quarter past nine in the evening _____ **f** Ten to eight in the morning _____

2 8.15 am to 10 am is 1 hour 45 minutes. How much time from

   a 10.30 am to 2.30 pm? _____   **b** 7.30 am to 5.15 pm? _____

   c 06.15 to 11.30? _____   **d** 09.35 to 13.55? _____

3 If you spend 5 minutes a day cleaning your teeth, how much is this in a year? _____

4 How many hours from 14.00 on Tuesday to 19.30 on Wednesday? _____

5 How many days from December 12th to January 9th inclusive? _____

6 London time is 5 hours ahead of New York time and 10 hours behind Sydney time.

Example (*i*) London time (03.30) → minus 5 hours → New York time (22.30)

Hours earlier than GMT | Hours later than GMT

11 10 9 8 7 6 5 4 3 2 1 GMT 1 2 3 4 5 6 7 8 9 10 11 12

Example (*ii*) London time (01.30) → plus 10 hours → Sydney time (11.30)

Now complete the following statements.

   a When it is 12.00 noon in London it is _____ in New York.

   b At 3.30 pm in Sydney it is _____ in London.

   c At 10.00 in New York it is _____ in London.

   d At 11.15 in London it is _____ in Sydney.

   e At 1.15 am in London it is _____ in New York and _____ in Sydney.

   f At _____ in New York it is 17.30 in London.

   g At _____ in Sydney it is 11.00 in London.

7 If I catch a flight at 14.15 in London and fly to New York I arrive there 6 hours later.
   What time will it be London time? _____
   What time will it be New York time? _____

8 If I fly from London to Sydney and leave at 08.00, what time will it be in Sydney when I arrive if the flight takes 23 hours? _____

TV Viewing

1 Here are the BBC and ITV schedules for a typical Monday.

BBC 1

*NOT IN COLOUR (R) REPEAT

9.0-11.25 SCHOOLS (and at 11.40-12.5 and 2.1-3.0). 11.25-11.40 **You and Me** (R). 12.45 **NEWS.**

1.0 PEBBLE MILL AT ONE including Plan Your Land. 1.45-2.1 **Fingerbobs** (R). 3.15 **Songs of Praise** from Winchester (R). 3.53 **Regional News** (except London).

3.55 PLAY SCHOOL 4.20 **Wally Gator** (R), 4.25 **Jackanory.** 4.40 **Battle of the Planets:** Cartoon series. 5.0 **John Craven's Newsround.** 5.5 **Blue Peter:** The Long Sleep – the tortoises go into hibernation. 5.35 **Paddington.** 5.40 **NEWS.**

5.55 NATIONWIDE (with regional variations). 6.55 **Angels:** Jay tries to keep on an even keel after her drinking problems.

7.20 THE ROCKFORD FILES: Black Mirror, part two. The beautiful Dr. Dougherty is being harassed by a mysterious assailant. Jim begins to suspect one of her patients.

8.10 PANORAMA. 9.0 NEWS.

9.25 POCKET MONEY (Film 1972).

11.0 FILM 79. 11.30 **News.** 11.32 **Roadshow Disco.** 11.57-12.5 **Weather, Regional News,** Closedown.

ITV

9.30 a.m. SCHOOLS. 12.0 **Jamie and the Magic Torch:** Animated adventure. 12.10 **Rainbow.** 12.30 **Emmerdale Farm** (R). 1.0 **NEWS.** 1.20 **Thames News.** 1.30 **All About Toddlers.** 2.0 **Heart to Heart.**

2.30 FOREIGN INTRIGUE (Film, 1955). 4.15 **Clapperboard:** Chris Kelly looks at the cinema scene. 4.45 **Why Can't I Go Home?:** Drama series about hospital life in a children's ward. 5.15 **Batman.** 5.45 **NEWS.**

6.0 THAMES AT 6: Regional news and views. 6.35 **Crossroads.**

7.0 GIVE US A CLUE: Michael Aspel hosts a new series of the male versus female charades game. Una Stubbs and Lionel Blair captain the teams.

7.30 CORONATION STREET.

8.0 ONLY WHEN I LAUGH: James Bolam, Peter Bowles star in a new comedy series set in a hospital.

8.30 THE MIGHTY MICRO: 1 – The Coming of the Micro-Processor. New series examining the implications of the silicon chip.

9.0 MINDER: Gunfight at the OK Launderette. Starring Dennis Waterman, George Cole. New film series following the fortunes of an ex-con and his suave boss – Terry "the Minder" and Arthur "his guv'nor." 10.0 **NEWS.**

10.30 THE GHOST OF FLIGHT 401: Starring Ernest Borgnine, Gary Lockwood. TV film about a flight engineer who finds himself living a nightmare after a malfunction on his aircraft. 12.20 Closedown.

a How long does *Panorama* last? _____

b How long does *The Ghost of Flight 401* last? _____

c How much longer is *Minder* than *The Rockford Files*? _____

2 a Programmes can be divided into the categories given in the table. Complete the table to show how much time is devoted to each category by the BBC and ITV.

	1 News and Current Affairs	2 Drama (Plays)	3 Films	4 Light Entertainment	5 Sport	6 Education & Children's	7 Serials
BBC			1 h 35 min		—		
ITV			3 h 35 min		—		

b Put the information in the table into a double bar chart.

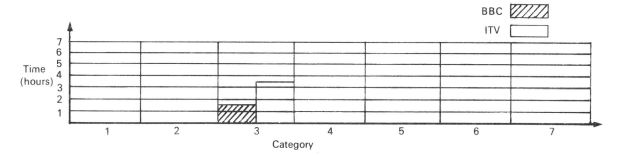

Speed and Time I

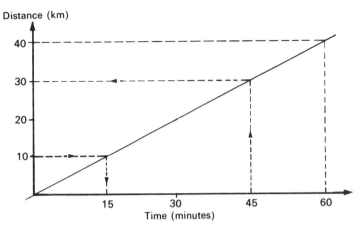

1 The graph shows the distance travelled by a car in 1 hour.

  a How far is this? _____

  b What speed is it travelling at?

  _____

  c To find out how far it goes in 45 minutes, we 'read off' the distance on the graph.
  It goes _____ km in 45 minutes.

  d To find out how long it takes to travel 10 km we 'read off' the other way.
  It takes _____ minutes to travel 10 km.

  e How long did it take to travel
  20 km? _____ minutes   35 km? _____ minutes   5 km? _____ minutes

  f How far did it travel in
  40 minutes? _____ km 10 minutes? _____ km $22\frac{1}{2}$ minutes? _____ km

2 A boy had to go 20 km to the nearest town to pick up a new bike. He got a lift in a car for the first 10 km which took 20 minutes. Then he took a bus for the next 10 km, and arrived 40 minutes later. He spent half an hour at the shop, and then rode the bike home. It took one and a quarter hours.

  a Complete the graph of the journey below.

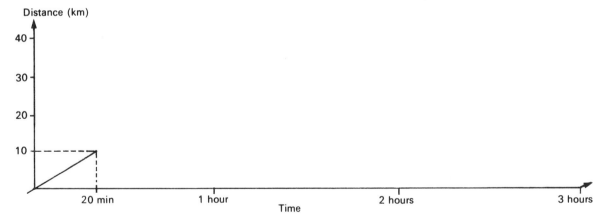

$$\text{Speed} = \frac{\text{Distance (km)}}{\text{Time (hours)}}$$

  b What was his average speed for the outward journey? _____

  c What was his average speed for the homeward journey? _____

  d Which journey was quicker, the outward or homeward? _____
  and by how much? _____

  e What was his average speed for the whole journey (including the stop)? _____

Speed and Time II

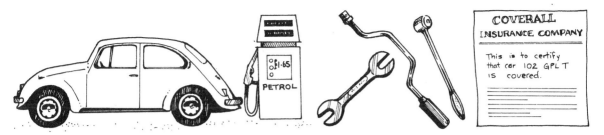

1 Mr Smith travels an average of 240 miles a week in his car.
He uses 8 gallons of petrol a week.

a How many miles does he do per gallon? _____

b If petrol is £1·65 per gallon, how much does he spend on petrol a week? _____

c How much is this a year? _____

d Complete this table of Mr Smith's car expenses.

Car expenses in a year	
Petrol	£
Tax	£50·00
Insurance	£110·00
Maintenance	£150·00
Depreciation	£200·00
Total	£

e How much on average does it cost Mr Smith to run his car each week? _____

f He has to have a new engine put in, and it costs £650·00.
What are his new weekly expenses? _____

g By how much are his weekly expenses increased? _____

2 This graph shows how many miles per gallon a car will do when it is being driven at various speeds.

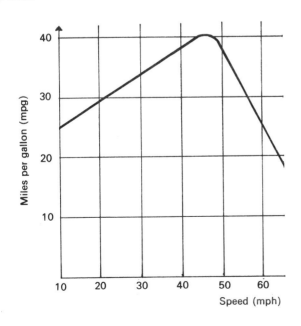

a How many miles to the gallon at 20 mph?

_____

b How many gallons are needed for a journey of 360 miles travelling at an average speed of 60 mph?

_____

c What is the most economical speed to drive at?

_____

Timetables I

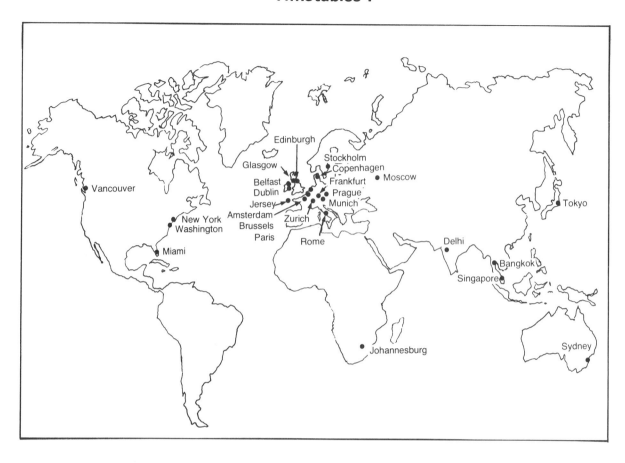

Here is a departure timetable for a busy British Airport.

TIME	AIRLINE	DESTINATION	TIME	AIRLINE	DESTINATION	TIME	AIRLINE	DESTINATION
12.00	British Airways	New York	12.20	Aeroflot	Moscow	12.40	Pan Am	Zurich
12.02	Laker	Stockholm	12.22	British Airways	Brussels	12.42	British Airways	Tokyo
12.04	Lufthansa	Frankfurt	12.24	Lufthansa	Munich	12.44	Air France	Paris
12.06	British Airways	Munich	12.26	Quantas	Sydney	12.46	Air Canada	Vancouver
12.08	K.L.M.	Amsterdam	12.28	Pan Am	New York	12.48	British Airways	Rome
12.10	Air Lingus	Belfast	12.30	British Airways	Miami	12.50	Air India	Delhi
12.12	S.A.A.	Johannesburg	12.32	Air Lingus	Dublin	12.52	Quantas	Singapore
12.14	British Airways	Paris	12.34	Pan Am	Washington	12.54	British Airways	Edinburgh
12.16	British Airways	Prague	12.36	British Airways	Jersey	12.56	Laker	New York
12.18	Brt. Caledonian	Glasgow	12.38	Dan Air	Copenhagen	12.58	Thai-International	Bangkok

1 How many are *British Airways* flights? _____

2 How many flights go to North America? _____

3 How many more flights do *British Airways* have than *Pan Am*? _____

4 How many flights are internal? _____

5 How many more flights go to Europe than to North America? _____

6 If you miss the 12.00 flight to New York, how long do you have to wait for the next one?

_____

7 How long is there between the two *Quantas* flights? _____

8 If the same number of flights leave every hour, how many would leave in a day? _____

9 What percentage of flights are *Pan Am*? _____

10 What percentage of flights go to Europe? _____

Timetables II

Here is a timetable for buses from Nottingham to London

	A1	A2	A3	A1	A2	A3
Nottingham	08.00	09.00	10.00	11.00	12.00	13.00
Loughborough		09.30	10.30			
Leicester		10.00	11.00			
Market Harboro'			11.30			
Northampton			12.30			
London	11.00	12.30	14.00			

1 How long does each bus take to travel from Nottingham to London?

    A1 _____   A2 _____   A3 _____

2 How much longer is the longest journey than the shortest journey? _____

3 What time does the 10.00 A3 get to Market Harboro'? _____

4 a If I catch a bus in Leicester at 11.00, what time will I get to London? _____
   b How long a journey is this? _____

5 a If I am in Nottingham at 09.30, how long do I have to wait for the next bus? _____
   b How long do I have to wait for the next direct bus? _____
   c What time will this direct bus arrive in London? _____

6 Fill in the timetable for the 12.00 A2

7 Fill in the timetable for the 13.00 A3

8 The last bus of the day is at six o'clock in the evening.
   a Will it be A1, A2, or A3? _____
   b What time will it get to London? _____

9 Mr Smith catches the 10.00 bus from Nottingham to Leicester. He breaks his journey there to see an aunt.
   a What time is the next bus from Leicester to London? _____
   b What time does it get to London? _____

Area

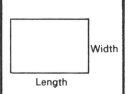

Rectangle

Area =
  length × width

A = l × w

Perimeter = 2l + 2w

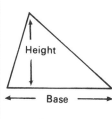

Triangle

Area = $\frac{1}{2}$ base × height

A = $\frac{1}{2}$ b × h

1 Find the areas of these shapes.

a A rectangle of length 5 cm, width 12 cm

Area = _____ cm²

b A rectangle of length 3 mm, width 8 mm

Area = _____ mm²

c A triangle of base 6 cm, height 4 cm

Area = _____ cm²

d A triangle of base 2 m, height 11m

Area = _____ m²

e A rectangle of length 7 cm, width 8$\frac{1}{2}$ cm

Area _____ cm²

f A triangle of base 5 cm, height 7 cm

Area = _____ cm²

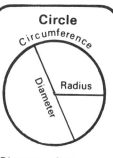

Circle

Diameter = 2 × radius

D = 2r

Circumference =
  π × diameter
($π = 3.1$ approx)
C = πd

Area =
  π × radius × radius
A = πr²

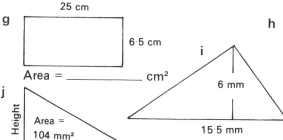

g 25 cm, 6.5 cm

Area = _____ cm²

h Area = 160m², 5 m

What is the length?

_____ m

i 6 mm, 15.5 mm

Area = _____ mm²

j Height, Area = 104 mm², 26 mm

What is the height? _____ mm

2 A circle has a radius of 4 cm.
Diameter = _____ cm. Circumference = _____ cm. Area = _____ cm².

3 A circle has a radius of 7 m.
Diameter = _____ m. Circumference = _____ m. Area = _____ m².

4

26 mm

Radius = _____ mm. Circumference = _____ mm.

Area = _____ mm².

5 A circle has a circumference of 68.2 cm.
Diameter = _____ cm. Radius = _____ cm. Area = _____ cm².

6

4.5 cm

Find the area of this circle.

_____ cm².

40

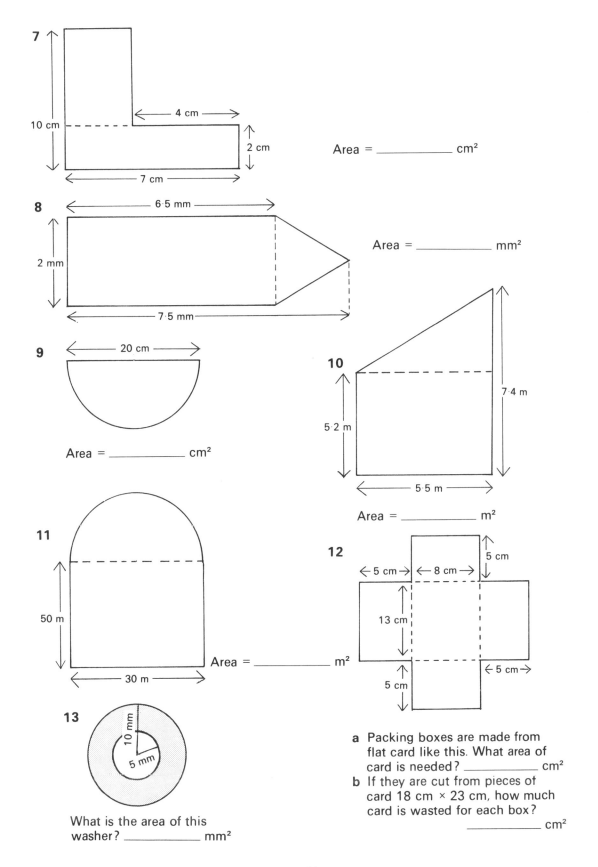

7 10 cm   4 cm   2 cm   7 cm

Area = _____ cm²

8 6·5 mm   2 mm   7·5 mm

Area = _____ mm²

9 20 cm

Area = _____ cm²

10 5·2 m   7·4 m   5·5 m

Area = _____ m²

11 50 m   30 m

Area = _____ m²

12 5 cm   8 cm   5 cm   13 cm   5 cm   5 cm

a Packing boxes are made from flat card like this. What area of card is needed? _____ cm²

b If they are cut from pieces of card 18 cm × 23 cm, how much card is wasted for each box?
_____ cm²

13 10 mm   5 mm

What is the area of this washer? _____ mm²

Volume

To find the volume of a shape, find the area of the front face then multiply by the depth.

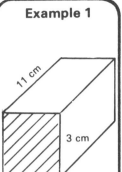
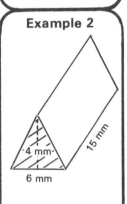
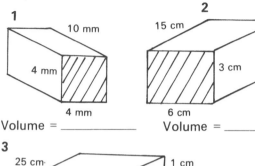

1

10 mm

4 mm

4 mm

Volume = _____

2

15 cm

3 cm

6 cm

Volume = _____

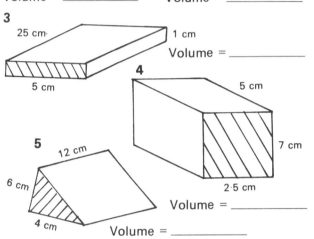

3

25 cm

1 cm

5 cm

Volume = _____

4

5 cm

7 cm

2·5 cm

Volume = _____

5

12 cm

6 cm

4 cm

Volume = _____

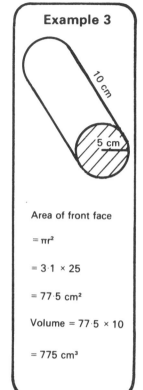
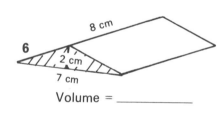

6

8 cm

2 cm

7 cm

Volume = _____

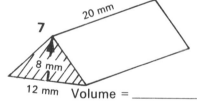

7

20 mm

8 mm

12 mm   Volume = _____

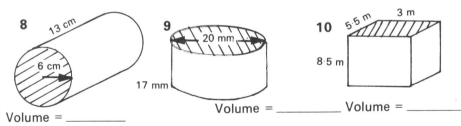

8

13 cm

6 cm

Volume = _____

9

20 mm

17 mm

Volume = _____

10

5·5 m   3 m

8·5 m

Volume = _____

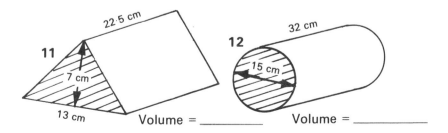

11

22·5 cm

7 cm

13 cm   Volume = _____

12

32 cm

15 cm

Volume = _____

42

13 A cuboid has length 8 cm, width 5 cm and the volume is 800 cm³. What is the depth?

_____ cm

14 How many packets of Crunchy Flakes 20 cm × 30 cm × 6 cm, can fit into a box 40 cm × 60 cm × 72 cm? _____

15 Here is a podium, used for Medal winners in the Olympic Games. What is its volume?

_____ m³

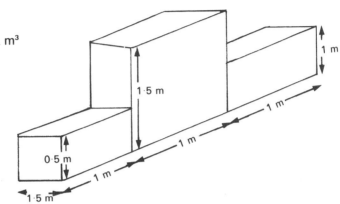

16 Here is a garden shed. What is its volume?

_____ m³

If planks of wood are 15 cm wide, how many metres are needed to make one side 2 m × 3 m?

_____

How many are needed to make the whole shed (excluding roof)?

_____

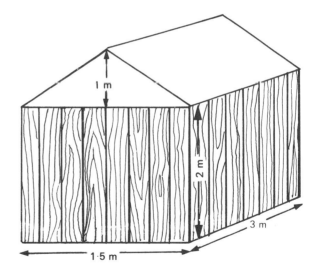

17 A path to a garage is 9 m long by 2·5 m wide. If I wish to concrete it 10 cm deep, what volume of concrete do I need? _____ m³

Scale Drawing and Maps I

Here is a scale drawing of a house plan. The scale is 1 cm to 1 m.

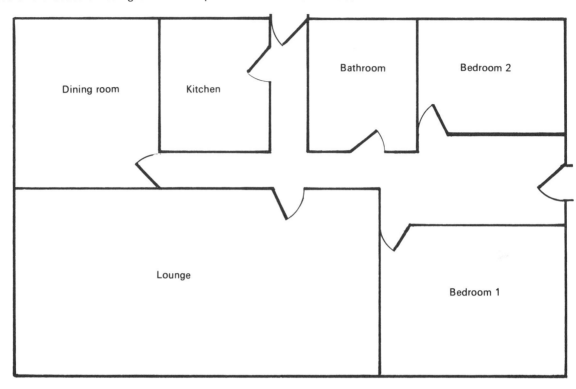

1 What is **a** the scale length _____, **b** the real length _____ of the house?

2 What is **a** the scale width _____, **b** the real width _____ of the house?

3 Complete this table by measuring up the rooms.

Room	Scale		Actual	
	Length	Width	Length	Width
Dining room	cm	cm	m	m
Kitchen	cm	cm	m	m
Bathroom	cm	cm	m	m
Bedroom 2	cm	cm	m	m
Bedroom 1	cm	cm	m	m
Lounge	10 cm	cm	10 m	m

4 The lounge has a stone fireplace 1·5 m wide.

   a Draw this on to the plan.

   b What is the scale width of the fireplace? _____

5 Each room has a window on the narrowest outside wall.
   Each window measures one-quarter of the wall's length.

   a Work out the actual length of each window.

   Dining room _____ m   Kitchen _____ cm   Bathroom _____ cm

   Bedroom 1 _____ m   Bedroom 2 _____ cm   Lounge _____ m

   b Draw the windows in scale on the plan.

6 What is the actual area of the hall? _____ m²

Scale Drawing and Maps II

Here is a simplified scale map of part of Britain's motorway complex.

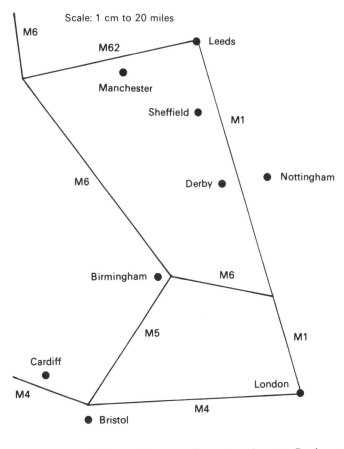

Scale: 1 cm to 20 miles

M6
M62
Leeds
Manchester
Sheffield
M1
M6
Derby
Nottingham
Birmingham
M6
M5
M1
Cardiff
London
M4
M4
Bristol

1 Measure the distance along the motorways to find approximately how far it is from

a London to Nottingham

_____ cm

This represents

_____ miles

b London to Bristol

_____ cm

This represents

_____ miles

c Leeds to Manchester

_____ cm

This represents

_____ miles

2 From Manchester to Preston is 2·5 cm. What distance does this represent? _____ miles

3 A commercial traveller goes from London, to Derby, to Sheffield, to Leeds, to Manchester, to Birmingham and then to Bristol and back to London in five days.

a Find the total distance for the round trip. _____ cm _____ miles

b How far does he travel, on average each day? _____ miles per day

c If he travels at an average of 50 miles an hour, what is the average length of time he spends travelling each day? _____

d His car does approximately 30 mpg.

How many gallons of petrol does he use? _____

4 a How far is a round trip from London to Birmingham to Bristol to London? _____

b Petrol is £1·65 a gallon, and a car does an average of 35 mpg.

How much does this round trip cost? _____

Scale Drawing and Maps III

1 Here is a deck plan of a boat **not** drawn to scale.

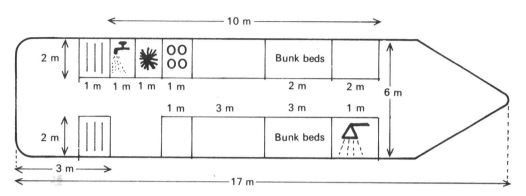

Re-draw the plan to the scale 1 cm = 2 m.

2 Here is a map of Britain.
Complete the table to show the distances between the places named.

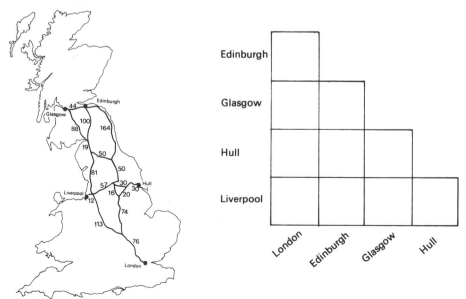

Decorating I

The Ketley family are going to decorate their lounge.
Here is a plan of the room.

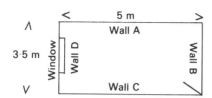

The walls are 3 m high.
The window is 2 m wide by 1 m high.
The door is 1 m wide by 2 m high.

Material	Price
Matt Vinyl Emulsion	£1·70 per litre
Gloss	£2·10 per litre
Wallpaper	£3·50 per roll

Material	Price
Ceiling Tex	75p per litre
Brushes	80p each
Wallpaper paste	25p per packet

1 Wall A

This wall is to be papered. Each roll of wallpaper is 50 cm wide and allowing for pattern, they can get 3 strips of room height from each roll.

a How many rolls will they need? _____

b What will the wallpaper cost? _____

c If they will need only one packet of wallpaper paste, what will be the total cost for Wall A? _____

2 Walls B, C and D

These are to be painted in Matt Vinyl Emulsion. A litre of Matt Vinyl Emulsion will cover 9 m².

a Area of Wall B = _____ m × _____ m = _____ m²

b Area of Wall C = (_____ m × _____ m) – area of door _____ m² = _____ m²

c Area of Wall D = (_____ m × _____ m) – area of window _____ m² = _____ m²

d Total area = _____ m² + _____ m² + _____ m² = _____ m²

e How many litres of Matt Vinyl Emulsion will they need? _____

f What will be the total cost for Walls B, C and D? _____

3 Ceiling

The ceiling is to be covered in Ceiling Tex. One litre of Ceiling Tex will cover 1·5 m².

a Area of the ceiling = _____ m × _____ m = _____ m²

b What will be the cost for the ceiling? _____

4 Door and Window

These are to be painted in gloss. If they will need one litre of gloss paint and two brushes, what will be the total cost for the door and window? _____

5 Floor

They intend to fit the floor with carpet costing £7·99 per m².

a What is the area of the floor? _____

b What will it cost for carpet? _____

6 Total Cost

What will the Ketleys pay to have their lounge decorated? _____

Decorating II

The Ketley's wish to decorate one of their bedrooms.
This is a plan of the room.

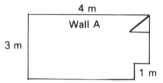

The walls are 3 m high.
The window is 1·3 m wide by 1 m high.
The door is 1 m wide by 2 m high.

Material	Price		Material	Price
Matt Vinyl Emulsion........	£1·70	per litre	Wallpaper paste........	25p per packet
Gloss paint..............	£2·10	per litre	Cork tiles............	50p each
Wallpaper..............	£2·50	per roll	Mirror tiles...........	65p each
Ceiling Tex..............	75p	per litre	Tube of glue..........	80p
Brushes................	80p	each	Roller Blind..........	£8·50

1 Wall A

Wall A is to be papered. Each roll of wallpaper is 50 cm wide and allowing for pattern, they can get 3 strips of room height from each roll.

a How many rolls will they need? _____

b What will the wallpaper cost? _____

2 The other walls

Somewhere on one of the other walls, they wish to have about 1 m² of mirror tiles. The mirror tiles measure 15 cm × 15 cm. The rest of the walls are to be painted in Matt Vinyl Emulsion. One litre covers 9 m².

a How many tiles will they need? _____

b What will the tiles cost? _____

c Remembering to deduct the window, door and mirror tile areas,
what area of wall is to be painted? _____

d How many litres of Matt Vinyl Emulsion will be needed? _____

e What will the Matt Vinyl Emulsion cost? _____

f What is the total cost for these other walls? _____

3 Ceiling

The ceiling is to be covered in Ceiling Tex. One litre of Ceiling Tex will cover 1·5 m².

a What area of ceiling is to be covered? _____

b How many litres are needed? _____

c What will the Ceiling Tex cost? _____

4 Floor

The floor is to be covered in cork tiles. Each tile measures 30 cm × 30 cm.

a How many tiles are needed? _____

b What will the cork tiles cost? _____

5 Extras

What will it cost for the following extras?

a 1 litre gloss paint _____ **b** 1 brush _____

c 1 packet of wallpaper paste _____ **d** 2 tubes of glue _____

e 1 roller blind _____

f What is the total cost of the extras? _____

6 Total Cost

What will be the total cost of decorating the bedroom? _____